Herbert Spencer

Spontaneous generation, and the hypothesis of

physiological units:

A reply to the North American review

Herbert Spencer

Spontaneous generation, and the hypothesis of physiological units:
A reply to the North American review

ISBN/EAN: 9783337716134

Printed in Europe, USA, Canada, Australia, Japan

Cover: Foto ©ninafisch / pixelio.de

More available books at **www.hansebooks.com**

SPONTANEOUS GENERATION,

AND THE HYPOTHESIS OF

PHYSIOLOGICAL UNITS.

A REPLY

TO

THE NORTH AMERICAN REVIEW.

BY

HERBERT SPENCER.

NEW YORK:

D. APPLETON AND COMPANY,

90, 92 & 94 GRAND STREET.

1870.

PUBLISHERS' NOTE.

The following letter was originally written for publication in the NORTH AMERICAN REVIEW; *but, in pursuance of a general rule not to admit replies, the editor declined it.* *To prevent misapprehension in regard to his views, however, Mr. Spencer designed to introduce it as an appendix to vol. 1 of the " Principles of Biology."* *A portion of it having been published in* APPLETONS' JOURNAL, *has awakened so much interest among men of science, that it has been thought desirable to issue it in this separate form.*

ON ALLEGED "SPONTANEOUS GENERATION," AND ON THE HYPOTHESIS OF PHYSIOLOGICAL UNITS.

The Editor of the North American Review.

SIR: It is in most cases unwise to notice adverse criticisms. Either they do not admit of answers, or the answers may be left to the penetration of readers. When, however, a critic's allegations touch the fundamental propositions of a book, and especially when they appear in a periodical having the position of the *North American Review*, the case is altered. For these reasons the article on "Philosophical Biology," published in your last number, demands from me an attention which ordinary criticisms do not.

It is the more needful for me to notice it, because its two leading objections have the one an actual fairness and the other an apparent fairness; and, in the absence of explanations from me, they will be considered as substantiated even by many, or perhaps most, of those who have read the work itself—much more by those who have not read it. That, to prevent the spread of misapprehensions, I ought to say something, is further shown by the fact that the same two objections have already been made in England—the one by Dr. Child, of Oxford, in his *Essays on Physiological Subjects*, and the other by a writer in the *Westminster Review* for July, 1865.

In the note to which your reviewer refers, I have, as he says, tacitly repudiated the belief in "spontaneous generation;" and that I have done this in such a way as to leave open the door for the interpretation given by him is true. Indeed, the fact that Dr. Child, whose criticism is a sympathetic one, puts the same construction on this note, proves that your reviewer has but drawn what seems to be a necessary inference. Nevertheless, the inference is one which I did not intend to be drawn.

In explanation, let me at the outset remark that I am placed at a disadvantage in having had to omit that part of the System of Philosophy which deals with Inorganic Evolution. In the original programme will be found a parenthetic reference to this omitted part, which should, as there stated, precede the *Principles of Biology*. Two volumes are missing. The closing chapter of the second, were it written, would deal with the evolution of organic matter—the step preceding the evolution of living forms. Habitually carrying with me in thought the contents of this unwritten chapter, I have, in some cases, expressed myself as though the reader had it before him; and have thus rendered some of my statements liable to misconstructions. Apart from this, however, the explanation of the apparent inconsistency is very simple, if not very obvious. In the first place, I do not believe in the "spontaneous generation" commonly alleged, and referred to in the note; and so little have I associated in thought this alleged "spontaneous generation" which I disbelieve, with the generation by evolution which I do believe, that the repudiation of the one never occurred to me as liable to be taken for repudiation of the other. That creatures having *quite specific structures* are evolved in the course of a few hours, without antecedents calculated to determine their specific forms, is to me incredible. Not only the established truths of Biology, but the established truths of science in general, negative the supposition that organisms, having structures definite enough to identify them as belonging to known genera and species, can be produced in the absence of germs derived from antecedent organisms of the same genera and species. If there can suddenly be imposed on simple protoplasm the organization which constitutes it a *Paramœcium*, I see no reason why animals of greater complexity, or indeed of any complexity, may not be constituted after the same manner. In brief, I do not accept these alleged facts as exemplifying Evolution, because they imply something immensely beyond that which Evolution, as I understand it, can achieve. In the second place, my disbelief extends not only to the alleged cases of "spontaneous generation," but to every case akin to them. The very conception of spontaneity is wholly incongruous with the conception of Evolution. For this reason I regard as objectionable Mr. Darwin's phrase "spontaneous variation" (as indeed he does himself); and I have sought to show that there are always assignable causes of variation. No form of Evolution, inorganic or organic, can be spontaneous; but in every instance the antecedent forces must be adequate in their quantities, kinds, and distributions, to work the observed effects. Neither the alleged cases of "spontaneous generation," nor any imaginable cases in the least allied to them, fulfil this requirement.

If, accepting these alleged cases of "spontaneous generation," I had assumed, as your reviewer seems to do, that the evolution of organic life commenced in an analogous way; then, indeed, I should have left myself open to a fatal criticism. This supposed "spontaneous generation" habitually occurs in menstrua that contain either organic matter, or matter originally derived from organisms; and such organic matter, proceeding in all known cases from organisms of a higher kind, implies the preëxistence of such higher organisms. By what kind of logic, then, is it inferrible that organic life was initiated after a manner like that in which *Infusoria* are said to be now spontaneously generated? Where, before life commenced, were the superior organisms from which these lowest organisms obtained their organic matter? Without doubting that there are those who, as the reviewer says, "can penetrate deeper than Mr. Spencer has done into the idea of universal evolution," and who, as he contends, prove this by accepting the doctrine of "spontaneous generation," I nevertheless think that I can penetrate deep enough to see that a tenable hypothesis respecting the origin of organic life must be reached by some other clew than that furnished by experiments on decoction of hay and extract of beef.

From what I do not believe, let me now pass to what I do believe. Granting that the formation of organic matter, and the evolution of life in its lowest forms, may go on under existing cosmical conditions; but believing it more likely that the formation of such matter and such forms took place at a time when the heat of the Earth's surface was falling through those ranges of temperature at which the higher organic compounds are unstable; I conceive that the moulding of such organic matter into the simplest types must have commenced with portions of protoplasm more minute, more indefinite, and more inconstant in their characters, than the lowest Rhizopods—less distinguishable from a mere fragment of albumen than even the *Protogenes* of Professor Haeckel. The evolution of specific shapes must, like all other organic evolution, have resulted from the actions and reactions between such incipient types and their environments, and the continued survival of those which happened to have specialties best fitted to the specialties of their environments. To reach by this process the comparatively well-specialized forms of ordinary *Infusoria*, must, I conceive, have taken an enormous period of time.

To prevent, as far as may be, future misapprehension, let me elaborate this conception so as to meet the particular objections raised. The reviewer takes for granted that a "first organism" must be assumed by me, as it is by himself. But the conception of a "first organism," in any thing like the current sense of the words, is wholly at variance with conception of evolution; and scarcely less at variance with the facts revealed by the micro- scope. The lowest living things are not, properly speaking, or- ganisms at all; for they have no distinctions of parts—no traces of organization. It is almost a misuse of language to call them "forms" of life: not only are their outlines, when distinguishable, too unspecific for description, but they change from moment to moment and are never twice alike, either in two individuals or in the same individual. Even the word "type" is applicable in but a loose way; for there is little constancy in their generic charac- ters: according as the surrounding conditions determine, they un- dergo transformations now of one kind and now of another. And the vagueness, the inconstancy, the want of appreciable struc- ture, displayed by the simplest of living things as we now see them, are characters (or absences of characters) which, on the hypothesis of Evolution, must have been still more decided when, as at first, no "forms," no "types," no "specific shapes," had been moulded. That "absolute commencement of organic life on the globe," which the reviewer says I "cannot evade the ad- mission of," I distinctly deny. The affirmation of universal evo- lution is in itself the negation of an "absolute commencement" of any thing. Construed in terms of evolution, every kind of being is conceived as a product of modifications wrought by in- sensible gradations on a preëxisting kind of being; and this holds as fully of the supposed "commencement of organic life" as of all subsequent developments of organic life. It is no more needful to suppose an "absolute commencement of organic life" or a "first organism," than it is needful to suppose an absolute commencement of social life and a first social organism. The as- sumption of such a necessity in this last case, made by early speculators with their theories of "social contracts" and the like, is disproved by the facts; and the facts, so far as they are ascer- tained, disprove the assumption of such a necessity in the first case. That organic matter was not produced all at once, but was reached through steps, we are well warranted in believing by the experiences of chemists. Organic matters are produced in the laboratory by what we may literally call *artificial evolution*. Chemists find themselves unable to form these complex combina- tions directly from their elements; but they succeed in forming them indirectly, by successive modifications of simpler combina- tions. In some binary compound, one element of which is pres- ent in several equivalents, a change is made by substituting for one of these equivalents an equivalent of some other element; so producing a ternary compound. Then another of the equivalents is replaced, and so on. For instance, beginning with ammonia, NH_3, a higher form is obtained by replacing one of the atoms of hydrogen by an atom of methyl, so producing methyl-amine, N $(C H_3 H_2)$; and then, under the further action of methyl, ending in a further substitution, there is reached the still more compound substance dimethyl-amine, N $(C H_3) (C H_3)$ H. And in this manner highly-complex substances are eventually built up. Another characteristic of their method is no less significant. Two complex compounds are employed to generate, by their action upon one another, a compound of still greater complexity: different heterogeneous molecules of one stage become parents of a molecule a stage higher in heterogeneity. Thus, having built up acetic acid out of its elements, and having, by the process of substitution described above, changed the acetic acid into propionic acid, and propionic into butyric, of which the formula is $\left\{ \begin{array}{l} C (C H_3) (C H_3) H \\ C O (H O) \end{array} \right\}$; this complex compound, by operating on another complex compound, such as the dimethyl-amine named above, generates one of still greater complexity, butyrate of dimethyl-amine $\left\{ \begin{array}{l} C (C H_3) (C H_3) H \\ C O (H O) \end{array} \right\}$ N $(C H_3) (C H_3)$ H.

See, then, the remarkable parallelism. The progress toward higher types of organic molecules is effected by modifications upon modifications; as throughout Evolution in general. Each of these modifications is a change of the molecule into equilibrium with its environment—an adaptation, as it were, to new surrounding conditions to which it is subjected; as throughout Evolution in general. Larger, or more integrated, aggregates (for compound molecules are such) are successively generated; as throughout Evolution in general. More complex or heterogeneous aggregates are so made to arise, one out of another; as throughout Evolution in general. A geometrically-increasing multitude of these larger and more complex aggregates so produced, at the same time results; as throughout Evolution in general. And it is by the action of the successively higher forms on one another, joined with the action of environing conditions, that the highest forms are reached; as throughout Evolution in general.

When we thus see the identity of method at the two extremes—when we see that the general laws of evolution, as they are exemplified in known organisms, have been unconsciously conformed to by chemists in the artificial evolution of organic matter; we can scarcely doubt that these laws were conformed to in the natural evolution of organic matter, and afterward in the evolution of the simplest organic forms. In the early world, as in the modern laboratory, inferior types of organic substances, by their mutual actions under fit conditions, evolved the superior types of organic substances, ending in organizable protoplasm. And it can hardly be doubted that the shaping of organizable protoplasm, which is a substance modifiable in multitudinous ways with extreme facility, went on after the same manner. As I learn from one of our first chemists, Prof. Frankland, *protein* is capable of existing under probably at least a thousand isomeric forms; and, as we shall presently see, it is capable of forming, with itself and other elements, substances yet more intricate in composition, that are practically infinite in their varieties of kind. Exposed to those innumerable modifications of conditions which the Earth's surface afforded, here in amount of light, there in amount of heat, and elsewhere in the mineral quality of its aqueous medium, this extremely changeable substance must have undergone now one, now another, of its countless metamorphoses. And to the mutual influences of its metamorphic forms under favoring conditions, we may ascribe the production of the still more composite, still more sensitive, still more variously-changeable portions of organic matter, which, in masses more minute and simpler than existing *Protozoa*, displayed actions verging little by little into those called vital—actions which protein itself exhibits in a certain degree, and which the lowest known living things exhibit only in a greater degree. Thus, setting out with inductions from the experiences of organic chemists at the one extreme, and with inductions from the observations of biologists at the other extreme, we are enabled deductively to bridge the interval—are enabled to conceive how organic compounds were evolved, and how, by a continuance of the process, the nascent life displayed in these became gradually more pronounced. And this it is which has to be explained, and which the alleged cases of " spontaneous generation " would not, were they substantiated, help us in the least to explain.

It is thus manifest, I think, that I have not fallen into the alleged inconsistency. Nevertheless, I admit that your reviewer was justified in inferring this inconsistency ; and I take blame to myself for not having seen that the statement, as I have left it, is open to misconstruction.

I pass now to the second allegation—that in ascribing to certain specific molecules, which I have called "physiological units," the aptitude to build themselves into the structure of the organism to which they are peculiar, I have abandoned my own principle, and have assumed something beyond the redistribution of Matter and Motion. As put by the reviewer, his case appears to be well made out; and that he is not altogether unwarranted in so putting it, may be admitted. Nevertheless, there does not in reality exist the supposed incongruity.

Before attempting to make clear the adequacy of the conception which I am said to have tacitly abandoned as insufficient, let me remove that excess of improbability the reviewer gives to it, by the extremely-restricted meaning with which he uses the word mechanical. In discussing a proposition of mine he says:

"He then cites certain remarks of Mr. Paget on the permanent effects wrought in the blood by the poison of scarlatina and small-pox, as justifying the belief that such a 'power' exists, and attributes the repair of a wasted tissue to 'forces analogous to those by which a crystal reproduces its lost apex.' (Neither of which phenomena, however, is explicable by mechanical causes.)"

Were it not for the deliberation with which this last statement is made, I should take it for a slip of the pen. As it is, however, I have no course left but to suppose the reviewer unaware of the fact that molecular actions of all kinds are now not only conceived as mechanical actions, but that calculations based on this conception of them bring out the results that correspond with observation. There is no kind of rearrangement among molecules (crystallization being one) which the modern physicist does not think of, and correctly reason upon, in terms of forces and motions like those of sensible masses. Polarity is regarded as a resultant of such forces and motions; and when, as happens in many cases, light changes the molecular structure of a crystal, and alters its polarity, it does this by impressing, in conformity with mechanical laws, new motions on the constituent molecules. That the reviewer should present the mechanical conception under so extremely limited a form, is the more surprising to me because, at the outset of the very work he reviews, I have, in various passages, based inferences on those immense extensions of it which he ignores; indicating, for example, the interpretation it yields of the inorganic chemical changes effected by heat, and the organic chemical changes effected by light (*Principles of Biology*, § 13).

Premising, then, that the ordinary idea of mechanical action must be greatly expanded, let us enter upon the question at issue—the sufficiency of the hypothesis that the structure of each organism is determined by the polarities of the special molecules, or physiological units, peculiar to it as a species, which necessitate tendencies toward special arrangements. My proposition and the reviewer's criticism upon it will be most conveniently presented if I quote in full a passage of his from which I have already extracted some expressions. He says:

"It will be noticed, however, that Mr. Spencer attributes the possession of these 'tendencies,' or 'proclivities,' to natural inheritance from ancestral organisms; and it may be argued that he thus saves the mechanist theory and his own consistency at the same time, inasmuch as he derives even the 'tendencies' themselves ultimately from the environment. To this we reply, that Mr. Spencer, who advocates the nebular hypothesis, cannot evade the admission of an absolute commencement of organic life on the globe, and that the 'formative tendencies,' without which he cannot explain the evolution of a single individual, could not have been inherited by the first organism. Besides, by his virtual denial of spontaneous generation, he denies that the first organism was evolved out of the inorganic world, and thus shuts himself off from the argument (otherwise plausible) that its 'tendencies' were ultimately derived from the environment."

This assertion is already in great measure disposed of by what has been said above. Holding that, though not " spontaneously generated," those minute portions of protoplasm which first displayed in the feeblest degree that changeability taken to imply life, were evolved, I am *not* debarred from the argument that the "tendencies" of the physiological units are derived from the inherited effects of environing actions. If the conception of a " first organism " were a necessary one, the reviewer's objection would be valid. If there were an "absolute commencement " of life, a definite line parting organic matter from the simplest living forms, I should be placed in the predicament he describes. But as the doctrine of Evolution itself tacitly negatives any such distinct separation; and as the negation is the more confirmed by the facts the more we know of them; I do not feel that I am entangled in the alleged difficulty. My reply might end here; but as the hypothesis in question is one not easily conceived, and very apt to be misunderstood, I will attempt a further elucidation of it.

Much evidence now conspires to show that molecules of the substances we call elementary are in reality compound; and that, by the combination of these with one another, and recombinations of the products, there are formed systems of systems of molecules, unimaginable in their complexity. Step by step, as the aggregate molecules so resulting grow larger and increase in heterogeneity, they become more unstable, more readily transformable by small forces, more capable of assuming various characters. Those composing organic matter transcend all others in size and intricacy of structure ; and in them these resulting traits reach their extreme. As implied by its name, *protein*, the essential substance of which organisms are built, is remarkable alike for the variety of its metamorphoses and the facility with which it undergoes them : it changes from one to another of its thousand isomeric forms on the slightest change of conditions. Now, there are facts warranting the belief that though these multitudinous isomeric forms of protein will not unite directly with one another, yet they admit of being linked together by other elements with which they combine. And it is very significant that there are habitually present two other elements, sulphur and phosphorus, which have quite special powers of holding together many equivalents—the one being pentatomic and the other hexatomic. So that it is a legitimate supposition (justified by analogies) that an atom of sulphur may be a bond of union among half a dozen different isomeric forms of protein ; and similarly with phosphorus. A moment's thought will show that, setting out with the thousand isomeric forms of protein, this makes possible a number of these combinations almost passing the power of figures to express. Molecules so produced, perhaps exceeding in size and complexity those of protein as those of protein exceed those of inorganic matter, may, I conceive, be the special units belonging to special kinds of organisms. By their constitution they must have a plasticity, or sensitiveness to modifying forces, far beyond that of protein; and bearing in mind not only that their varieties are practically infinite in number, but that closely allied forms of them, chemically indifferent to one another as they must be, may coexist in the same aggregate, we shall see that they are fitted for entering into unlimited varieties of organic structures.

The existence of such physiological units, peculiar to each species of organism, is not unaccounted for. They are evolved simultaneously with the evolution of the organisms they compose —they differentiate as fast as these organisms differentiate; and are made multitudinous in kind by the same actions which make the organism they compose multitudinous in kind. This conception is clearly representable in terms of the mechanical hypothesis. Every physicist will indorse the proposition that in each aggregate there tends to establish itself an equilibrium between the forces exercised by all the units upon each and by each upon all. Even in masses of substance so rigid as iron and glass, there goes on a molecular rearrangement, slow or rapid according as circumstances facilitate, which ends only when there is a complete balance between the actions of the parts on the whole and the actions of the whole on the parts: the implication being that every change in the form or size of the whole necessitates some redistribution of the parts. And though, in cases like these, there occurs only a polar rearrangement of the molecules, without changes in the molecules themselves; yet where, as often happens, there is a passage from the colloid to the crystalloid state, a change of constitution occurs in the molecules themselves. These truths are not limited to inorganic matter: they unquestionably hold of organic matter. As certainly as molecules of alum have a form of equilibrium, the octahedron, into which they fall when the temperature of their solvent allows them to aggregate, so certainly must organic molecules of each kind, no matter how complex, have a form of equilibrium in which, when they aggregate, their complex forces are balanced—a form far less rigid and definite, for the reason that they have far less definite polarities, are far more unstable, and have their tendencies more easily modified by environing conditions. Equally certain is it that the special molecules, having a special organic structure as their form of equilibrium, must be reacted upon by the total forces of this organic structure; and that, if environing actions lead to any change in this organic structure, these special mole-

cules, or physiological units, subject to a changed distribution of
the total forces acting upon them will undergo modification—
modification which their extreme plasticity will render easy.   By
this action and reaction I conceive the physiological units peculiar to each kind of organism to have been moulded along with
the organism itself.   Setting out with the stage in which protein
in minute aggregates took on those simplest differentiations
which fitted it for differently-conditioned parts of its medium,
there must have unceasingly gone on perpetual readjustments
of balance between aggregates and their units—actions and reactions of the two, in which the units tended ever to establish the
typical form produced by actions and reactions in all antecedent
generations, while the aggregate, if changed in form by change
of surrounding conditions, tended ever to impress on the units a
corresponding change of polarity, causing them in the next generation to reproduce the changed form—their new form of equilibrium.

   This is the conception which I have sought to convey, though
it seems unsuccessfully, in the *Principles of Biology ;* and which
I have there used to interpret the many involved and mysterious
phenomena of Genesis, Heredity, and Variation.   In one respect
only am I conscious of having so inadequately explained myself
as to give occasion for a misinterpretation—the one made by the
Westminster reviewer above referred to.   By him, as by your
own critic, it is alleged that in the idea of " inherent tendencies "
I have introduced, under a disguise, the conception of " the archæus, vital principle, *nisus formativus,* and so on."   This allegation is in part answered by the foregoing explanation.   That
which I have here to add, and did not adequately explain in the
Principles of Biology, is, that the proclivity of units of each
order toward the specific arrangement seen in the organism they
form is not to be understood as resulting from their own structures and actions only, but as the product of these and the environing forces to which they are exposed.   Organic evolution
takes place only on condition that the masses of protoplasm
formed of the physiological units, and of the assimilable materials out of which others like themselves are to be multiplied, are
subject to heat of a given degree—are subject, that is, to the un-
ceasing impacts of undulations of a certain strength and period ;
and, within limits, the rapidity with which the physiological
units pass from their indefinite arrangement to the definite arrangement they presently assume is proportionate to the strengths
of the ethereal undulations falling upon them.   In its complete
form, then, the conception is that these specific molecules, having
the immense complexity above described, and having correspond-
ently complex polarities which cannot be mutually balanced by
any simple form of aggregation, have, for the form of aggregation

in which all their forces are equilibrated, the structure of the adult organism to which they belong; and that they are compelled to fall into this structure by the coöperation of the environing forces acting on them, and the forces they exercise on one another—the environing forces being the source of the *power* which effects the rearrangement, and the polarities of the molecules determining the *direction* in which that power is turned. Into this conception there enters no trace of the hypothesis of an "archæus or vital principle;" and the principles of molecular physics fully justify it.

It is, however, objected that "the living body in its development presents a long succession of *differing* forms; a continued series of changes for the whole length of which, according to Mr. Spencer's hypothesis, the physiological units must have an 'inherent tendency.' Could we more truly say of any thing, 'it is unrepresentable in thought?'" I reply that if there is taken into account an element here overlooked, the process will not be found "unrepresentable in thought." This is the element of size or mass. To satisfy or balance the polarities of each order of physiological units, not only a certain structure of organism, but a certain size of organism is needed; for the complexities of that adult structure in which the physiological units are equilibrated cannot be represented within the small bulk of the embryo. In many minute organisms, where the whole mass of physiological units required for the structure is present, the very thing *does* take place which it is above implied *ought* to take place. The mass builds itself directly into the complete form. This is so with *Acari*, and among the nematoid *Entozoa*. But among higher animals such direct transformations cannot happen. The mass of physiological units required to produce the size as well as the structure that approximately equilibrates them is not all present, but has to be formed by successive additions—additions which in viviparous animals are made by absorbing, and transforming into these special molecules, the organizable materials directly supplied by the parent, and which in oviparous animals are made by doing the like with the organizable materials in the "food-yelk," deposited by the parent in the same envelope with the germ. Hence it results that, under such conditions, the physiological units which first aggregate into the rudiment of the future organism do not form a structure like that of the adult organism, which, when of such small dimensions, does not equilibrate them. They distribute themselves so as partly to satisfy the chief among their complex polarities. The vaguely-differentiated mass thus produced cannot, however, be in equilibrium. Each increment of physiological units formed and integrated by it changes the distribution of forces; and this has a double effect. It tends to modify the differentiations already made, bringing them a step nearer to the equilibrating structure; and the physiological units next integrated, being brought under the aggregate of polar forces exercised by the whole mass, which now approaches a step nearer to that ultimate distribution of polar forces which exists in the adult organism, are coerced more directly into the typical structure. Thus there is necessitated a series of compromises. Each successive form assumed is unstable and transitional: approach to the typical structure going on hand in hand with approach to the typical bulk.

Possibly I have not succeeded by this explanation, any more than by the original explanation, in making this process "representable in thought." It is manifestly untrue, however, that I have, as alleged, reintroduced under a disguise the conception of a "vital principle." That I interpret embryonic development in terms of Matter and Motion, cannot, I think, be questioned. Whether the interpretation is adequate, must be a matter of opinion; but it is clearly a matter of fact that I have not fallen into the inconsistency asserted by your reviewer. At the same time I willingly admit that, in the absence of certain statements which I have now supplied, he was not unwarranted in representing my conception in the way that he has done.

But, while I consider that what your reviewer has said on these two essential points falls within the limits of legitimate criticism, I do not consider that he is justified in much that he says by implication respecting my general views.

In the first place, he conveys a totally wrong idea of the mode of interpretation he criticises. He gives his readers no conception of the immense extensions which modern science has made of the "mechanical theory," now applied to the solution of all physical phenomena whatever; but he has deliberately restricted its applications in a way that produces an appearance of difficulty where no difficulty exists. The common uses of the words "mechanical" and "mechanist" are such as inevitably call up in all minds the notions of visible masses of matter acting on one another by measurable forces and producing sensible motions. In the absence of explanations or illustrations serving to enlarge the conception thus suggested, so as to bring within it the oscillations of the molecules of matter, and the undulations of the molecules of ether pervading all space, even the cultivated reader must carry with him an extremely crude and narrow idea of the "mechanist theory," and cannot fail to be struck with the seeming absurdity of interpreting vital phenomena in mechanical terms. But the reviewer says nothing to prevent misconceptions so arising. He gives no hint that heat, light, and electricity, are now all recognized as "modes of motion;" and that most of their phenomena are mechanically interpreted, while the rest are regarded as mechanically interpretable. He does not explain that the " mechanist " theory in its comprehensive form embraces actions such as those by which variations in the solar spots cause variations in our magnetic needles, and actions such as those through which Sirius tells us what substances are contained in his atmosphere. True, he makes a passing reference to chemical changes as being included by me under the conception of mechanical; but he leaves this as a dead statement quite unintelligible to the general reader; and in the typical example he gives of my mode of interpretation (the development of vertebræ by transverse strains) he deliberately excludes the physio-chemical and chemical actions which I imply as coöperating, and describes me as attributing the effects entirely to the pressures and tensions caused by muscular movements! (See p. 408.) Instead of the developed ideas of Matter and Motion everywhere implied throughout the *Principles of Biology*, the reviewer leads every one to suppose that I bring to bear on biological problems nothing beyond the vulgar ideas of Matter and Motion, and leaves me responsible for the ludicrous incongruity!

That, however, which I regard as most reprehensible in his criticism is the way in which he persists in representing the *System of Philosophy* I am working out as a materialistic system. Already he has once before so represented it, and the injustice of so representing it has been pointed out. He knows that I have repeatedly and emphatically asserted that our conceptions of Matter and Motion are but symbols of an Unknowable Reality; that this Reality cannot be that which we symbolize it to be; and that, as manifested beyond consciousness under the forms of Matter and Motion, it is the same as that which, in consciousness, is manifested as Feeling and Thought. Yet he continues to describe me as reducing every thing to dead mechanism. If his statement on pp. 383–4 has any meanig at all, it means that there exists some " force operating *ab extra*," some " external power " distinguished by him as " mechanical," which is not included in that immanent force of which the universe is a manifestation; though whence it comes he does not tell us. This conception he speaks of as though it were mine; making it seem that I ascribe the moulding of organisms to the action of this " mechanical " " external power," which is distinct from the Inscrutable Cause of things. Yet he either knows, or has ample means of knowing, that I deny every such second cause : indeed, he has himself classed me as an opponent of dualism. I recognize no forces within the organism, or without the organism, but the variously-conditioned modes of the universal immanent force; and the whole process of organic evolution is everywhere attributed by me to the coöperation of its variously-conditioned modes, internal and external. That this has been all along my general view, is clearly shown in the closing paragraph of *First Principles*, where I have said—

"A Power of which the nature remains forever inconceivable, and to which no limits in Time or Space can be imagined, works in us certain effects. These effects have certain likenesses of kind, the most general of which we class together under the names of Matter, Motion, and Force; and between these effects there are likenesses of connection, the most constant of which we class as laws of the highest certainity. Analysis reduces these several kinds of effect to one kind of effect; and these several kinds of uniformity to one kind of uniformity. And the highest achievement of Science is the interpretation of all orders of phenomena, as differently-conditioned manifestations of this one kind of effect, under differently-conditioned modes of this one kind of uniformity. But, when Science has done this, it has done nothing more than systematize our experience; and has in no degree extended the limits of our experience. We can say no more than before, whether the uniformities are as absolutely necessary, as they have become to our thoughts relatively necessary. The utmost possibility for us, is an interpretation of the process of things as it presents itself to our limited consciousness; but, how this process is related to the actual process, we are unable to conceive, much less to know. Similarly, it must be remembered that, while the connection between the phenomenal order and the ontological order is forever inscrutable, so is the connection between the conditioned forms of being and the unconditioned form of being forever inscrutable. The interpretation of all phenomena in terms of Matter, Motion, and Force, is nothing more than the reduction of our complex symbols of thought to the simplest symbols; and when the equation has been brought to its lowest terms the symbols remain symbols still. Hence the reasonings contained in the foregoing pages afford no support to either of the antagonist hypotheses respecting the ultimate nature of things. Their implications are no more materialistic than they are spiritualistic; and no more spiritualistic than they are materialistic. Any argument which is apparently furnished to either hypothesis is neutralized by as good an argument furnished to the other. The Materialist, seeing it to be a necessary deduction from the law of correlation that what exists in consciousness under the form of feeling is transformable into an equivalent of mechanical motion, and by consequence into equivalents of all the other forces which matter exhibits, may consider it therefore demonstrated that the phenomena of consciousness are material phenomena. But the Spiritualist, setting out with the same data, may argue with equal cogency, that if the forces displayed by matter are cognizable only under the shape of those equivalent amounts of consciousness which they produce, it is to be inferred that these forces, when existing out of consciousness, are of the same intrinsic nature as when existing in consciousness; and that so is justified the spiritualistic conception of the external world, as consisting of something essentially identical with what we call mind. Manifestly, the establishment of correlation and equivalence between the forces of the outer and the inner worlds may be used to assimilate either to the other; according as we set out with one or other term. But he who rightly interprets the doctrine contained in this work will see that neither of these terms can be taken as ultimate. He will see that though the relation of subject and object renders necessary to us these antithetical conceptions of Spirit and Matter, the one is no less than the other to be regarded as but a sign of the Unknown Reality which underlies both."

This is the conception which your reviewer continues to speak of as "mechanical" and "mechanist," without giving his readers any suspicion of the qualified sense in which only these words can be applied. If he thinks that by doing this he has represented the conception with fairness, or with any approach to fairness, I cannot agree with him.

I am, sir, yours, etc.,
HERBERT SPENCER.

SPENCER'S

SYSTEM OF PHILOSOPHY.

THE attention of the reading public is invited to this great work of Mr. Herbert Spencer, now in course of regular republication in this country.

When, in 1860, he issued the prospectus of a comprehensive philosophical scheme in which it was proposed to embrace the subjects of Life, Mind, Society, and Morality in one system of scientific principles, the project was naturally looked upon by many as chimerical. The term "Philosophy" had been so long applied to speculations upon remote and transcendental questions yielding only empty and discordant results, that a strong prejudice existed against any thing calling itself by that name; so that the announcement of a New Philosophical System was looked upon, not as promising a contribution to real knowledge, but only as a fresh display of futile ingenuity in an exhausted and a hopeless field. It was, besides, thought impossible to reduce such diverse subjects to any thing like a philosophical unity of treatment; while the task, even if possible, was held to be beyond the power of any single mind to accomplish.

Uninfluenced by these considerations, Mr. Spencer went forward with his undertaking; and that he has not entered upon the pursuit of a chimera, but upon the solution of a problem which, though vast, is still practicable, will be apparent when we consider his point of view. By "Philosophy" he understands the deepest explanation of the universal order at which man can arrive. Renouncing all hope of passing beyond phenomena and their laws, he proposes, as the task of Philosophy, the determination of those broadest principles which are disclosed within the phenomenal order of the universe. The several sciences give us separate explanations of the divisions of Nature; Philosophy seeks by combining these explanations to arrive at universal principles. It is the tendency of all science to disclose a unity in the method of Nature; Philosophy thus becomes an expression of that unity, and may hence be defined as the highest unification of knowledge. Here is a definite, legitimate, and fruitful field of inquiry.

The largest principle of action in the world of natural phenomena Mr. Spencer found shadowed forth as an onward movement from one order of things to another—a kind of vague, progressive unfolding. The latest astronomy, for example, declares that the solar system has been evolved from a nebulous origin. So geology affirms that the earth was once a molten mass of chaotic material. We are familiar with the growth of plants from seeds; of animals from ova; and with the decay and disappearance of living forms. Mind also has its unfolding and its decline; while the arts of civilization and the changes of social life have been progressive. But, although vague analogies may have been obscurely perceived among those remote and widely-diverse phenomena, nothing like a common and clearly-defined law, binding them together, was at all suspected. The first step in this direction was taken by the eminent naturalist Von Baer early in the century. He obtained a partial glimpse of the ultimate principle of growth in the world of life, and formulated his conception into a law of organic development. Mr. Spencer took up the inquiry at this point. He showed that Von Baer's conception of the principle was incomplete, and he worked it into greater fulness and clearness, determining its elements, limitations, and causes. He discovered also that the law of organic development, when analyzed and defined with precision, turned out to be equally applicable to all orders of phenomena, and was in fact nothing less than a universal Law of Evolution.

The view thus arrived at had a new and startling significance. The conception of a universe in evolution — of astronomic systems, of our planet, of the spheres of life, mind, and society, governed in all their changes by one demonstrable rule of action, created an epoch in the advance of thought. A new point of departure was reached: the highest problems of human inquiry were reset in new lights and new relations, and a new Philosophy of the world became not only possible, but inevitable. Mr. Spencer grasped the full import of the situation, and entered unhesitatingly upon the great task of working out the new views into a coherent and all-comprehensive system of thought. Personally the circumstances were unpropitious. With very imperfect health, without means, and without a publisher to relieve him of business responsibilities, he closed with this formidable undertaking, and, in carrying it on under these crushing difficulties, he has exhibited a moral intrepidity in admirable keeping with the intellectual grandeur of the task.

The development of the system is now considerably advanced. We have the foundation volume, "First Principles," devoted to an elucidation of the Law of Evolution. Next follows the "Principles of Biology," in two volumes, in which that law is systematically applied to the phenomena of life. Based upon the Biology is the third division, the "Principles of Psychology," of which two completed parts, the Data of Psychology, and the Inductions of Psychology, have been lately republished. Thus a sufficiently large portion of the scheme is finished to enable us to judge of its claims; and we do not hesitate to say that, in the whole range of scientific and philosophic literature, nothing more clear, comprehensive, original, and profound, has yet appeared than the parts of Spencer's Philosophy which are now before the world. The most advanced thinkers in the leading countries of Europe have recognized its claims, and justify the declaration that it is beyond doubt the most important intellectual enterprise of our time. If this statement is objected to, as only an individual opinion, it is easily reënforced by authorities that will not be questioned.

As an indication of the position which Mr. Spencer has attained on the Continent, it may be mentioned that all his works are now being translated into Russian and published at St. Petersburg, by M. Thieblin, and that three translators, all Professors of Philosophy, are occupied in rendering them into French. Dr. Cazelles has translated " First Principles," and is now engaged upon the " Biology;" Prof. Ribot, of the *Lycée Imperial*, Laval, is translating the " Principles of Psychology; " and Prof. Péthore, of the *Lycée Imperial*, Angoulême, has translated the " Education" and the "Classification of the Sciences."

4

It may be further stated that Mr. Spencer's philosophical works have been adopted as text-books for the science classes in the University of Oxford, and questions upon them set for examination of students.

We cannot here offer any recent opinions of the English press on Mr. Spencer's *System of Philosophy*, for the sufficient reason that he has of late sent out no copies of his work for review.* The misstatements made by critics, who were either too idle to read his books or incompetent to form estimates of them, he had from the beginning taken as things to be expected and endured. Not long ago, however, he found, from personal testimony, that even those most interested in some of the topics treated by him had, for years, been deterred from looking at his works by the false impressions which press-notices had given to them. On learning this, he concluded that it would be best no longer to give occasion for these misleading criticisms. Neither of his last two volumes has been issued to the newspapers or the literary journals, and his English publishers have now a standing order not to send out for review any future work of his.

But, though no indorsement by anonymous critics can be cited, we may cite indorsements of much greater weight; the publicly-expressed opinions of well-known men, thoroughly competent to judge. The reader's attention is asked to the following extracts from works published by authors whose names are given :

---

* Nevertheless, incidental references constantly occur in the leading periodicals, which show the position Mr. Spencer has won in the world of thought. A late number of the *Saturday Review*, for example, has the following: "If we were to give our own judgment, we should say that, since Newton, there has not in England been a philosopher of more remarkable speculative and systematizing talent than (in spite of some errors and some narrowness) Mr. Herbert Spencer."—*London Saturday Review, February* 6, 1869.

Again, an able writer in the last number of the *Westminster Review*, in criticising Lecky's "History of Morals," thus speaks of Mr. Spencer: "Before entering on the discussion of his views, we cannot refrain from offering our tribute of respect to one who, whether for the extent of his positive knowledge, or for the profundity of his speculative insight, has already achieved a name second to none in the whole range of English philosophy, and whose works will worthily sustain the credit of English thought in the present generation."

Again, referring to Mr. Spencer's application of the Principles of Evolution to ethical philosophy, the writer says: "It appears to us that, even in its present stage, this theory of the progressive development of moral sentiments in the hereditary conscience of the race is the greatest advance which has been made in ethereal speculation since the time of Hartley."— *Westminster Review, October,* 1869.

5

Prof. Masson, *of the University of Edinburgh.*

"No such defect can be charged against the other writer whom I am now to name—Mr. Herbert Spencer. Of all our thinkers he is the one who, as it appears to me, has formed to himself the largest new scheme of a systematic philosophy, and, in relation to some of the greatest questions of Philosophy in their most recent forms, as set or reset by the last speculations and revelations of science, has already shot his thoughts the farthest. He both works out his Philosophy physiologically and psychologically from the centre, and—what seems to me an eminent merit in relation to the intellectual needs of the time—surveys it and contemplates it from the circumference cosmologically. Indeed, I should say that he is the British thinker who has most distinctly realized the absolute necessity that Philosophy lies under, of dealing with the total cosmological conception as well as with the mere psychical or physiological organism (and this from the demonstrable inter-relatedness of both), if it would grasp all the present throbbings of the speculative intellect. His writings take for granted this necessity, and make it plainer than it would otherwise be. Nowhere else are the various sciences so fished for generalizations that may come together as a whole to help in forming a Philosophy. Nowhere else, at all events, is there a more beautiful and fearless exposition of some of those recent scientific notions which I spoke of in the last chapter as affecting our views of metaphysical problems. There are parts of Mr. Spencer's writings, occupied with such expositions, which, from sheer scientific clearness, and adequacy of language to the matter, have all the effect of a poem. If even only for such renderings of high scientific conceptions, on the chance of their somehow taking possession of the popular soul, and uniting there to rectify previous forms of thought, he would deserve honorable recognition.

"But Mr. Spencer does not stop short in the character of an interpreter between Science and Philosophy, handing on the conceptions of Science to that congress of all the Powers, where they are to be adjusted and take effect. He assumes the work of the philosopher proper. He seeks to enmesh the physical round of things, as Science now orbs it to the instructed imagination, within a competent Metaphysic; he desires to fix in the centre a competent Psychology, consistent with this Metaphysic, and yet empirically and physiologically educed; and he would fill up the interior, or what of it the physical sciences leave void, with a competent Ethics, a competent Jurisprudence, a competent Æsthetics, a competent Science of Government and Politics.

"In this great work he is still engaged: and it will not perhaps be till the whole is accomplished that there will be the means of determining either the sufficiency of Mr. Spencer's philosophy for the higher practical purposes of philosophy, or its exact intellectual relations to previous systems. Already, in consequence both of the decisiveness of his views and the variety of interesting subjects over which they extend, Mr. Spencer, more than any other systematic British thinker save Mill, has an avowed following both here and in America; and, if any individual influence is visibly encroaching on Mill's in this country, it is his. For my own part, believing that no type of man ought to be more precious to a nation than a resolute systematic thinker, and believing Mr. Spencer to be a very high specimen of this type, I anticipate nothing but good—nothing, at least, but a clearing away of the bad—from what he has already done, or may yet do."—*Recent British Philosophy*, Chapter iv.

Mr. John Stuart Mill.

"The last extract is from Mr. Herbert Spencer, whose Principles of Psychology, in spite of some doctrines which he holds in common with the intuitive school, are, on the whole, one of the finest examples we possess of the Psychological method in its full power."—Mill's Hamilton, Chapter xiii.—"One of the acutest Metaphysicians of modern times."—*Ibid.*, Chapter ii.

"Mr. Spencer is one of the small number of persons who, by the solidity and

encyclopædical character of their knowledge, and their power of coördination
and concatenation, may claim to be peers of M. Comte, and entitled to a vote in
the estimate of him."—*Mill's Review of Comte.*—"One of the most vigorous as
well as boldest thinkers that English speculation has yet produced."—*Ibid.*

M. LAUGEL.

"The great work on Philosophy by Herbert Spencer, whom I would style
the last of English Metaphysicians. In the midst of universal indifference, Mr.
Spencer remained steadily attached to his philosophical studies, displaying all
that heroic courage and that rare independence indispensable to those who de-
vote themselves to toilsome researches which at best only recompense the stu-
dent with a few obscure and isolated suffrages. If Mr. Spencer, with his talents,
his fertility of genius, and the almost encyclopædic variety of knowledge of
which his writings furnish the proof, had chosen to follow the beaten path, nothing would have been more easy than for him to secure all those honors of which
English society is so prodigal to those who serve her as she wishes to be served.
He preferred, however, with a noble and touching self-denial, to put up with pov-
erty, and, what is still more difficult, with obscurity. But he deserves more than
vain assurances of sympathy: we must not merely admire his fidelity to profit-
less studies; his work itself merits the individual attention of all friends of Philosophy."—*Revue des Deux Mondes, of February* 15, 1864.

Prof. T. H. HUXLEY, F. R. S., LL. D.

"Those who hold the doctrine of Evolution (and I am one of them) conceive that there are grounds for believing that the world, with all that is in it
and on it, did not come into existence in the condition in which we now see it,
nor in any thing approaching that condition.

"On the contrary, they hold that the present conformation and composition
of the earth's crust, the distribution of land and water, and the infinitely diversi-
fied forms of animals and plants which constitute its present population, are
merely the final terms in an immense series of changes which have been brought
about, in the course of immeasurable time, by the operation of causes more or
less familiar to those which are at work at the present day.

"Perhaps this doctrine of Evolution is not maintained consciously and in its
logical integrity by a very great number of persons. The only complete and
systematic statement of the doctrine with which I am acquainted is that contained in Mr. Herbert Spencer's 'System of Philosophy ;' a work which should
be carefully studied by all who desire to know whither scientific thought is tend-
ing."—*Lecture before the Royal Institute of Great Britain.*

JAMES McCOSH, D. D., LL. D.

"In the new edition of his 'Intuitions of Mind,' Dr. McCosh devotes a chapter to Spencer's Philosophy. But, 'great as are the author's intellectual powers,' he believes that its success is beyond the reach of this 'giant mind,' and
indeed beyond the present possibility of science. Yet he observes : 'This in
regard to his theory as a whole ; but his bold generalizations are always sugges-
tive, and some may in the end be established as the profoundest laws of the
knowable universe.' "

JOSEPH D. HOOKER, F. R. S., LL. D.

"Another instance of successful experiment in Physiological Botany is Mr.
Herbert Spencer's observations on the circulation of the sap and the formation
of wood in plants. As is well known, the tissues of our herbs, shrubs, and trees,
from the tips of their roots to those of their petals and pistils, are permeated by
tubular vessels. The functions of these have been hotly disputed, some physi-
ologists affirming that they convey air, others fluids, others gases, and still others
assigning to them far-fetched uses, of a wholly different nature. By a series of
admirably contrived and conducted experiments, Mr. Spencer has shown not

7

only that these vessels are charged at certain seasons of the year with fluid, but that they are intimately connected with the formation of wood. He further investigates the nature of the special tissues concerned in this operation, and shows not merely how they may act, but to a great extent how they do act. As this paper will, I believe, be specially alluded to by the President of the Biological Section, I need dwell no further on it here, than to quote it as an example of what may be done by an acute observer and experimentalist, versed in physics and chemistry, but, above all, thoroughly instructed in scientific methods."— *Inaugural Address at Meeting of British Association*, 1868.

"One of our deepest thinkers, Mr. Herbert Spencer."—*Ibid.*

GEORGE HENRY LEWES.

"This last-named writer (Mr. Spencer) is now daily rising into wider influence. . . . . Even antagonists are compelled to admit the force and clearness of his genius, and the extent and profundity of his scientific knowledge. It is questionable whether any thinker of finer calibre has appeared in our country, although the future alone is to determine the position he is to assume in history. . . . . He alone, of all British thinkers, has organized a philosophy."— *History of Philosophy*, vol. ii., p. 653.

J. D. MORRELL, LL. D.

"Among modern English psychologists the author to whom I have been most indebted in this work is Mr. Herbert Spencer; more especially to the very able analysis which he has given of the process of reasoning in its qualitative and quantitative forms. So far as I have touched upon the theory of reasoning at all, I have followed to a large extent the pathway that he has pointed out, and which appears to me the most successful analysis which this subject has yet received in our own country."—*Outlines of Mental Philosophy.*

"Mr. Spencer is equally remarkable for his search after first principles; for his acute attempts to decompose mental phenomena into their primary elements; and for his broad generalizations of mental activity, viewed in connection with Nature, instinct, and all the analogies presented by life in its universal aspects." —*Medico-Chirurgical Review.*

DR. FRANCIS WAYLAND, *late President of Brown University.*

"I have read Herbert Spencer through, and some of the Essays twice. His volume on Education will do much to change the opinions of the civilized world. I hope it will be widely read here and in England. As to the worth of knowledge, he is very strong; here he and I are aiming at the same thing. I did not expect to see in my day one with whose views I could so sincerely sympathize. He speaks to the common-sense of humanity, and hates sham; and he will triumph, though it will take some time first."—*Life of Dr. Wayland*, vol. ii., p. 294.

J. G. MACVICAR, D. D.

"An author who is both extensively and profoundly versed in science, and who writes, on all the subjects which he handles, with great power, equally of observation, abstraction, and generalization: for such is Herbert Spencer, author of a 'System of Philosophy,' now in course of publication."—*Mind: its Powers and Capacities*, p. 9.

MR. GEORGE RIPLEY.

The most striking feature of Mr. Spencer as a philosopher, is the consummate degree in which he unites the power of abstract thought with the knowledge of scientific conclusions. He is as keen an analyst as is known in the history of Philosophy: I do not except either Aristotle or Kant, whom he greatly resembles in the trait to which I allude. . . . . He is familiar with the whole domain of the universe, whether that of consciousness or of observation. His outsight, so to speak, is as keen as his insight. No abstraction is too delicate and subtle for his mental grasp."—*Letter to the New York Tribune.*

WORKS OF HERBERT SPENCER,

PUBLISHED BY

D. APPLETON AND COMPANY.

SYSTEM OF PHILOSOPHY.

I.—FIRST PRINCIPLES.

(New and Enlarged Edition.)

PART I.—THE UNKNOWABLE.
PART II.—LAWS OF THE KNOWABLE.
559 pages. Price,   -   -   -   -   -   -   -   -   **$2.50**

II.—THE PRINCIPLES OF BIOLOGY.—VOL. I.

PART I.—THE DATA OF BIOLOGY.
PART II.—THE INDUCTIONS OF BIOLOGY.
PART III.—THE EVOLUTION OF LIFE.
475 pages. Price,   -   -   -   -   -   -   -   -   **$2.50**

PRINCIPLES OF BIOLOGY.—VOL. II.

PART IV.—MORPHOLOGICAL DEVELOPMENT.
PART V.—PHYSIOLOGICAL DEVELOPMENT.
PART VI.—LAWS OF MULTIPLICATION.
565 pages. Price,   -   -   -   -   -   -   -   -   **$2.50**

III.—THE PRINCIPLES OF PSYCHOLOGY.

PART I.—THE DATA OF PSYCHOLOGY.
144 pages. Price,   -   -   -   -   -   -   -   -   **$0.75**
PART II.—THE INDUCTIONS OF PSYCHOLOGY.
146 pages. Price,   -   -   -   -   -   -   -   -   **$0.75**

MISCELLANEOUS.

I.—ILLUSTRATIONS OF UNIVERSAL PROGRESS.

THIRTEEN ARTICLES. 451 pages. Price,   -   -   -   -   **$2.50**

II.—ESSAYS:

MORAL, POLITICAL, AND ÆSTHETIC.

TEN ESSAYS. 386 pages. Price,   -   -   -   -   **$2.50**

III.—SOCIAL STATICS:

OR THE CONDITIONS ESSENTIAL TO HUMAN HAPPINESS SPECIFIED, AND THE
FIRST OF THEM DEVELOPED.
523 pages. Price,   -   -   -   -   -   -   -   **$2.50**

IV.—EDUCATION:

INTELLECTUAL, MORAL, AND PHYSICAL.
283 pages. Price,   -   -   -   -   -   -   -   **$1.75**

V.—CLASSIFICATION OF THE SCIENCES.

50 pages. Price,   -   -   -   -   -   -   -   **$0.25**

VI.—SPONTANEOUS GENERATION, &c.

16 pages. Price, ...   **$0.25**

www.ingramcontent.com/pod-product-compliance
Lightning Source LLC
Chambersburg PA
CBHW031156090426
42738CB00008B/1357